# CARING FOR MARMOSET MONKEY

A COMPLETE GUIDE TO MARMOSET MONKEY HABITAT, DIET, PROS AND CONS, MANAGEMENT, AND MANY MORE INCLIUDED

**DR HUNTER DAVIS**

# Table of Contents

# CHAPTER ONE

## Introduction

### 1.1 Background

M armoset monkeys, belonging to the Callitrichidae family, are small New World primates known for their unique features and social behaviors. This section provides an overview of the historical and biological background of these fascinating creatures.

### 1.1.1 Evolutionary Origins

- Exploration of the evolutionary history and origins of marmosets within the primate order.
- Insights into their divergence from other primate species and adaptations over time.

### 1.1.2 Natural Habitats

Examination of the geographical distribution of marmosets in the wild.

Description of the diverse habitats they inhabit, ranging from tropical rainforests to savannas.

### 1.1.3 Taxonomic Context

- Overview of the taxonomic classification of marmosets, including their genus and species.
- Brief comparison to other primate families for contextual understanding.

### 1.1.4 Significance in Research

- Discussion of the role of marmosets in scientific research and their contributions to various fields.
- Highlighting key studies and discoveries that have involved marmoset monkeys.

### 1.1.5 Cultural and Historical Connections

- Exploration of marmosets' presence in indigenous cultures and historical contexts.
- Recognition of their symbolic importance and any historical interactions with human societies.

### 1.1.6 Captive Populations

- Examination of the history of marmosets in captivity, including their roles in zoos and research institutions.
- Consideration of challenges and benefits associated with captive marmoset populations.

### 1.2 Purpose of the work

The purpose of this work is to provide a thorough and informative exploration of Marmoset Monkeys, covering various aspects of their biology, behavior, conservation status, and interactions with humans. This section outlines the specific objectives and goals that this work aims to achieve.

### 1.2.1 Educational Resource

- To serve as an educational tool for individuals interested in learning about Marmoset Monkeys.
- Offering insights into their evolutionary history, taxonomy, and physical characteristics.

### 1.2.2 Conservation Awareness

- Raising awareness about the conservation status of marmosets and the challenges they face in the wild.
- Highlighting the importance of conservation efforts to ensure the survival of these primate species.

### 1.2.3 Behavior and Social Dynamics

- Exploring the intricate social structures and behaviors of marmoset monkeys in both wild and captive settings.
- Providing a deeper understanding of their communication methods and reproductive strategies.

### 1.2.4 Ethical Considerations in Human-Marmoset Interaction

- Examining the ethical implications of keeping marmosets as pets and their role in human entertainment.

- Encouraging responsible and compassionate interactions between humans and marmosets.

### 1.2.5 Scientific Significance

- Discussing the contributions of marmosets to scientific research, particularly in fields such as neuroscience and medicine.
- Offering insights into the unique characteristics that make them valuable subjects for scientific studies.

### 1.2.6 Cultural and Historical Context

- Exploring the historical connections and cultural significance of marmosets, if any.
- Recognizing their portrayal in art, literature, and indigenous traditions.

### 1.3 Scope of the Content

This section outlines the boundaries and coverage of the work concerning Marmoset Monkeys. It defines the specific areas and topics that will be addressed while acknowledging the limitations of the information presented.

### 1.3.1 Biological Focus

- In-depth exploration of the biological aspects, including taxonomy, physical characteristics, and evolutionary history of marmosets.

- Examination of their natural habitats and the adaptations that contribute to their survival in diverse environments.

### 1.3.2 Behavioral Insights
- Thorough analysis of the social structures, communication methods, and reproductive behaviors of marmoset monkeys.
- Comparison of behaviors between different species and the impact of captivity on their natural behaviors.

### 1.3.3 Conservation Considerations
- Comprehensive coverage of the conservation status of marmosets, including the threats they face in the wild.
- Highlighting ongoing conservation efforts and initiatives aimed at protecting and preserving marmoset populations.

### 1.3.4 Human-Marmoset Interaction
- Exploration of the ethical considerations surrounding the keeping of marmosets as pets and their involvement in human entertainment.
- Discussion of guidelines for responsible interactions and potential challenges in managing human-marmoset relationships.

### 1.3.5 Scientific Significance and Research Applications

- Investigation into the scientific contributions of marmosets, particularly in fields like neuroscience and medical research.
- Discussion of the challenges and advancements in using marmosets as model organisms in scientific studies.

1.3.6 Cultural and Historical Context (if applicable)
- Exploration of any cultural or historical connections of marmosets, including their portrayal in art, literature, and indigenous traditions.
- Recognition of their symbolic significance and any historical interactions with human societies.

While this work aims to provide a comprehensive overview of Marmoset Monkeys, it acknowledges that certain aspects may have limited information or ongoing research. The scope is designed to offer a well-rounded understanding within the available knowledge and resources up to the given cutoff date.

# CHAPTER TWO

## Taxonomy and Classification

## 2.1 Scientific Classification

Scientific classification, also known as taxonomy, is a hierarchical system used to categorize and organize living organisms based on shared characteristics and evolutionary relationships. The scientific classification of Marmoset Monkeys involves placing them within various taxonomic ranks, from broader categories down to more specific ones.

### 2.1.1 Kingdom: Animalia

Marmoset Monkeys, like all animals, belong to the Kingdom Animalia. This classification signifies that they are multicellular, eukaryotic organisms capable of mobility and obtaining nourishment through ingestion.

### 2.1.2 Phylum: Chordata

Within the Animalia Kingdom, Marmosets fall under the Phylum Chordata. This phylum includes animals possessing a notochord, a dorsal nerve cord, pharyngeal slits, and a post-anal tail at some stage in their development.

### 2.1.3 Subphylum: Vertebrata

Marmosets further belong to the Subphylum Vertebrata, which encompasses animals possessing a vertebral column or spine. This characteristic distinguishes vertebrates from other chordates.

### 2.1.4 Class: Mammalia

Marmosets are classified under the Class Mammalia, indicating that they are warm-blooded, have hair or fur on their bodies, and nurse their young with mammary glands. This class includes a diverse range of mammals, from rodents to whales.

### 2.1.5 Order: Primates

Within the Class Mammalia, Marmoset Monkeys are placed in the Order Primates. This order includes lemurs, tarsiers, monkeys, apes, and humans. Primates are characterized by features such as forward-facing eyes, grasping hands and feet, and large brains relative to body size.

### 2.1.6 Family: Callitrichidae

The family Callitrichidae is specific to Marmoset Monkeys. This family includes small-sized New World monkeys with distinct features, such as claws instead of nails on some fingers, a dental formula that differs from other primates, and the presence of twin births as a common reproductive strategy.

2.1.7 Genus: Callithrix, Callibella, Mico, and Cebuella
Marmosets are further divided into different genera.
Common genera include Callithrix, Callibella, Mico, and
Cebuella. Each genus contains specific species of
marmosets with unique characteristics and traits.

2.1.8 Species
The species classification is the most specific level in the
scientific classification hierarchy. Different species of
marmosets are identified within each genus, such as
Callithrix jacchus (Common Marmoset) and Callibella
humilis (Roosmalens' Dwarf Marmoset).

In summary, the scientific classification of Marmoset
Monkeys reflects their evolutionary relationships and
shared characteristics with other organisms. The
hierarchical structure allows scientists and researchers to
systematically organize and study different species,
facilitating a better understanding of the diversity and
complexity of the animal kingdom.

2.2 Common Species of Marmoset Monkeys
Marmoset Monkeys comprise several species, each with
its own unique characteristics and distribution. Here,
we'll delve into some common species, highlighting their
distinguishing features and habitats.

### 2.2.1 Common Marmoset (Callithrix jacchus)

Physical Characteristics:

- Small size: Common Marmosets typically weigh around 300 to 400 grams.
- Coat color: Fur coloration varies but is often a mix of brown, gray, and white.
- Distinctive features: White tufts of hair around the ears and a long, furry tail.

Habitat and Distribution:

Native to northeastern Brazil, they inhabit a range of environments, including tropical forests and scrublands. Highly adaptable, they can thrive in disturbed habitats and are often found near human settlements.

Behavior and Social Structure:

- Social animals: Common Marmosets live in family groups, exhibiting cooperative behaviors in foraging and raising offspring.
- Vocal communication: They use a variety of vocalizations for communication within the group.

### 2.2.2 Pygmy Marmoset (Cebuella pygmaea)

Physical Characteristics:

Smallest primate: Pygmy Marmosets are among the tiniest primates, weighing around 120 to 140 grams.
Fur coloration: Typically gray or brown, with dark markings on the face and a long, non-prehensile tail.

Habitat and Distribution:

- Native to the western Amazon Basin, including parts of Peru, Brazil, Ecuador, and Colombia.
- Prefers dense vegetation and is often found in rainforests.

Behavior and Social Structure:

- Arboreal lifestyle: Pygmy Marmosets are highly adapted to life in the trees, using their sharp claws for gripping branches.
- Family groups: They live in small family units, and like other marmosets, they often give birth to twins.

2.2.3 Goeldi's Marmoset (Callimico goeldii)
**Physical Characteristics:**
- Unique appearance: Goeldi's Marmoset has a distinct black fur coat and lacks the white ear tufts seen in other marmoset species.

- Facial features: Dark face with white markings around the mouth and forehead.

Habitat and Distribution:

- Native to the upper Amazon Basin, including parts of Peru, Brazil, and Colombia.
- Inhabits both primary and secondary forests.

Behavior and Social Structure:

- Social structure: Goeldi's Marmoset lives in small family groups, often with a monogamous pair.
- Dietary adaptation: Unlike other marmosets, they have a specialized diet that includes a significant portion of tree exudates.

These common species showcase the diversity within the marmoset family, each adapted to its specific environment and displaying unique social and behavioral characteristics. Studying these species contributes to a broader understanding of primate evolution and ecology.

## 2.3 Distribution and Habitat

## 2.3 Distribution and Habitat of Marmoset Monkeys

Marmoset monkeys are found in various regions of Central and South America, and their distribution is closely tied to specific habitats that provide the necessary resources for their survival. Here, we'll explore the distribution and habitat characteristics of marmosets, considering the common species mentioned earlier.

2.3.1 Common Marmoset (Callithrix jacchus)
Distribution:

- Native to northeastern Brazil, the common marmoset has been introduced to other regions, including parts of North and Central America.
- Populations can be found in urban areas, making them highly adaptable to disturbed habitats.

Habitat:

**Tropical and subtropical forests:** Common marmosets inhabit a variety of forest types, ranging from primary rainforests to secondary growth and scrublands.
**Edge habitats**: They are often observed in edge environments, where the forest meets open areas or human settlements.

Behavioral Adaptations:

- **Versatile diet**: Common marmosets are omnivores, feeding on fruits, insects, tree sap, and small vertebrates.
- Arboreal lifestyle: They are highly skilled climbers, utilizing their prehensile tails and specialized hands for navigating through the forest canopy.

2.3.2 Pygmy Marmoset (Cebuella pygmaea)
Distribution:

Indigenous to the western Amazon Basin, including parts of Peru, Brazil, Ecuador, and Colombia.
Occupies a range of habitats within the Amazon rainforest.

Habitat:

- Canopy dwellers: Pygmy marmosets primarily inhabit the upper canopy of tropical rainforests, utilizing their small size and agility to navigate branches.
- Dense vegetation: They prefer areas with dense vegetation, where they can easily find food and seek refuge from predators.

Behavioral Adaptations:

**Specialized diet:** Pygmy marmosets feed on tree sap, gum, fruit, and small insects.

Small group structure: They live in family units and exhibit cooperative behaviors in foraging and caring for offspring.

2.3.3 Goeldi's Marmoset (Callimico goeldii)
Distribution:

- Native to the upper Amazon Basin, spanning parts of Peru, Brazil, and Colombia.
- Inhabits both primary and secondary forests.

Habitat:

Diverse forest environments: Goeldi's marmosets are found in various forest types, from pristine primary rainforests to areas that have undergone some level of disturbance.

Riparian zones: They are often associated with areas near water, such as riverbanks.
Behavioral Adaptations:

- Unique diet: Goeldi's marmosets have a specialized diet that includes a significant portion of tree exudates, such as sap and gum.
- Small family groups: They live in family units, and monogamous pairs are commonly observed.

Understanding the distribution and habitat preferences of marmosets is crucial for conservation efforts and contributes to our knowledge of their ecological roles within their respective ecosystems. It also highlights the adaptability of certain species to changing landscapes, including human-altered environments.

# CHAPTER THREE

## Physical Characteristics

3.1 Size and Weight

3.1 Size and Weight of Marmoset Monkeys
Marmoset monkeys are small-sized primates known for their diminutive stature and lightweight build. The size and weight characteristics vary among different species, but collectively, they share common features that distinguish them from larger primates. Here, we'll explore the size and weight of marmosets, with a focus on the general trends and specific details for the common species mentioned earlier.

3.1.1 General Size Characteristics
Small Stature:

- Marmosets are among the smallest primates, exhibiting a diminutive stature compared to many other monkeys and apes.
- Their small size allows them to navigate the complex environment of the forest canopy with agility.

Average Body Length:

- The body length of marmosets varies among species but generally ranges from 5 to 12 inches (13 to 30 cm) without considering the tail.
- The tail, which is often prehensile, can add an additional 6 to 15 inches (15 to 38 cm) to their overall length.

3.1.2 Common Marmoset (Callithrix jacchus)
Weight Range:

Common marmosets typically weigh between 300 to 400 grams (10.5 to 14 ounces).
Females are generally slightly smaller and lighter than males.

Body Length:

- Their body length (excluding the tail) ranges from approximately 7 to 12 inches (18 to 30 cm).
- The tail can add an additional 8 to 15 inches (20 to 38 cm) to their total length.

3.1.3 Pygmy Marmoset (Cebuella pygmaea)
Weight Range:

- Pygmy marmosets are the smallest primates, weighing around 120 to 140 grams (4 to 5 ounces).

- Their lightweight build makes them adapted for an arboreal lifestyle.

Body Length:

- The body length of pygmy marmosets ranges from 4.6 to 6.2 inches (12 to 16 cm), excluding the tail.
- The tail can measure between 6.7 to 9.1 inches (17 to 23 cm).

3.1.4 Goeldi's Marmoset (Callimico goeldii)
Weight Range:

Goeldi's marmosets are slightly larger than pygmy marmosets, with weights ranging from 400 to 600 grams (14 to 21 ounces).

Body Length:

- Their body length (excluding the tail) is approximately 8 to 9 inches (20 to 23 cm).
- The tail can add an additional 11 to 14 inches (28 to 36 cm) to their total length.

3.1.5 Size Adaptations and Importance
Arboreal Adaptations:

The small size of marmosets is an adaptation to their arboreal (tree-dwelling) lifestyle, allowing them to move easily through the forest canopy.

Energetic Efficiency:

The lightweight build of marmosets is advantageous for conserving energy, especially during activities such as climbing and leaping between branches.

Reproductive Strategies:
The small size of marmosets is reflected in their reproductive strategies, with many species giving birth to twins. The relatively small size of infants facilitates easier care by the parents.

Understanding the size and weight characteristics of marmosets provides insights into their adaptations for an arboreal existence and their ecological niche within the tropical and subtropical environments they inhabit.

**Fur and Coloration**

3.2 Fur and Coloration of Marmoset Monkeys
The fur and coloration of marmoset monkeys are distinctive features that contribute to their visual

appearance and adaptability in their natural habitats. Each species may exhibit unique patterns and colors, and these characteristics often play a role in social signaling, camouflage, and thermoregulation. Let's explore the fur and coloration of marmosets, focusing on the common species mentioned earlier.

3.2.1 Common Marmoset (Callithrix jacchus)
Fur Characteristics:

- The fur of common marmosets is dense and soft, providing insulation and protection in various environmental conditions.
- They have a characteristic mane of white fur surrounding their face, extending to the sides of their head.

Coloration:

Fur coloration varies, but common marmosets typically have a mix of brown, gray, and white fur.
The dorsal (back) side often exhibits darker hues, while the ventral (belly) side tends to be lighter in color.

Facial Markings:

Dark facial markings around the eyes and nose contribute to their distinctive appearance.

The presence of white ear tufts is a notable feature.

### 3.2.2 Pygmy Marmoset (Cebuella pygmaea)
Fur Characteristics:

The fur of pygmy marmosets is dense, providing insulation in their canopy-dwelling habitat.
Their fur is short, soft, and may have a slightly coarse texture.

Coloration:

Fur coloration varies from gray to brown, with darker markings on the face.
The tail is often ringed with alternating dark and light bands.

Camouflage Adaptation:

The coloration helps pygmy marmosets blend into the dappled sunlight and shadows of the forest canopy, providing a form of natural camouflage.

### 3.2.3 Goeldi's Marmoset (Callimico goeldii)
Fur Characteristics:

The fur of Goeldi's marmoset is dense and black, contributing to its unique appearance.

The fur is short and velvety, giving them a distinct tactile quality.

Coloration:

Black fur covers their body, face, and tail.
White markings around the mouth and forehead provide contrast against the dark fur.

Adaptation to Riparian Zones:

The black coloration may serve as a form of camouflage in the shadows of the forest understory, especially in riparian zones near water.

### 3.2.4 Significance of Fur and Coloration
Social Signaling:

Fur coloration and facial markings play a role in social signaling within marmoset groups.
Contrasting colors may help individuals recognize each other within the complex environment of the forest.

Thermoregulation:

The dense fur of marmosets aids in thermoregulation, providing insulation against temperature variations in their habitat.

Adaptation to Habitat:

The specific coloration of each marmoset species is often adapted to its natural habitat, providing advantages in terms of survival, communication, and reproductive success.

Understanding the fur and coloration of marmoset monkeys enhances our appreciation for the adaptations that have evolved in response to their ecological niches and social structures. These features are not only aesthetically interesting but also crucial for their survival in diverse environments.

**Tail Characteristics**

3.3 Tail Characteristics of Marmoset Monkeys

The tail is a distinctive and functionally important feature for marmoset monkeys, contributing to their arboreal lifestyle, social interactions, and overall adaptation to their environments. Different species of marmosets exhibit variations in tail characteristics. Let's explore the tail characteristics of the common marmoset (Callithrix jacchus), pygmy marmoset (Cebuella pygmaea), and Goeldi's marmoset (Callimico goeldii).

### 3.3.1 Common Marmoset (Callithrix jacchus)
Prehensile Tail:

Common marmosets have a long and prehensile tail, meaning it is adapted for grasping and holding objects.
The tail is used for stability and balance while navigating through the branches of trees.
Length and Appearance:

- The tail of the common marmoset is approximately 8 to 15 inches (20 to 38 cm) long.
- It is covered in fur and can be used as a fifth limb to assist in climbing and leaping.

Social Signaling:

The tail is often involved in social interactions, such as grooming and maintaining proximity within the family group.

### 3.3.2 Pygmy Marmoset (Cebuella pygmaea)
Non-Prehensile Tail:

Pygmy marmosets have a long, slender, and non-prehensile tail.
While not adapted for grasping like the common marmoset's tail, it aids in balance during arboreal movements.

Ringed Appearance:

- The tail of the pygmy marmoset is often ringed with alternating dark and light bands.
- This ringed appearance is a distinctive feature of their tail.

Role in Locomotion:

The tail is important for maintaining balance and stability, especially when leaping between branches in the forest canopy.

3.3.3 Goeldi's Marmoset (Callimico goeldii)
Tail Characteristics:
- Goeldi's marmoset has a relatively long tail, which is not prehensile.
- The tail is covered in fur and contributes to their overall appearance.

Adaptation to Habitat:

In their riparian habitat, the tail likely assists in maintaining balance while navigating through the vegetation and along branches.

Limited Role in Social Interaction:

While the tail may play a role in balance, it might have a more limited role in social interactions compared to prehensile-tailed species.

### 3.3.4 Importance of Tail Characteristics

Arboreal Adaptation:

The tail of marmosets, whether prehensile or non-prehensile, is crucial for their arboreal lifestyle, aiding in climbing, leaping, and maintaining balance in the treetops.

Balance and Locomotion:

The tail serves as a dynamic tool for maintaining stability during rapid and agile movements, essential for their survival in the complex canopy environment.

Social Communication:

In prehensile-tailed marmosets, the tail often plays a role in social signaling and interactions within family groups.

Understanding the specific tail characteristics of different marmoset species provides insights into their ecological adaptations and behavioral strategies in their respective habitats. The tail is a versatile and multifunctional appendage that contributes significantly to their overall agility and survival in arboreal environments.

# CHAPTER FOUR

## Behavior and Social Structure

### Social Groups

4.1 Social Groups of Marmoset Monkeys

Marmoset monkeys are highly social creatures, and their social structures play a crucial role in their daily lives, communication, and reproductive strategies. Understanding the dynamics of marmoset social groups provides insights into their cooperative behaviors, communication methods, and the ways in which they navigate their environments. Let's explore the social groups of marmoset monkeys, with a focus on the common marmoset (Callithrix jacchus).

4.1.1 Common Marmoset (Callithrix jacchus)

Family Structure:

- Common marmosets typically live in family groups, also known as troops or bands.
- A family group consists of an adult breeding pair and their offspring.

Monogamous Pairing:

Marmosets exhibit monogamous pairing, where a single male and female form a long-term breeding partnership. This monogamous bond is crucial for cooperative behaviors, including parenting and foraging.

Cooperative Breeding:

In common marmoset family groups, cooperative breeding is a notable feature. Older siblings, and occasionally other group members, assist in caring for the infants.
Cooperative breeding enhances the survival and well-being of the offspring.

Communication within Groups:

- Marmosets are known for their complex vocal communication. They use a variety of calls and vocalizations to convey information within the family group.
- Communication aids in coordinating group activities, such as foraging and responding to potential threats.

4.1.2 Social Bonds and Grooming
Strong Social Bonds:

Marmosets form strong social bonds within their family groups.
Social grooming, where individuals groom each other's fur, reinforces social bonds and contributes to group cohesion.

Scent-Marking:
Scent-marking is another social behavior observed in marmoset groups. Glands on the chest and genital regions produce scent secretions used for marking territories and identifying individuals.

4.1.3 Reproductive Behavior
Twin Births:

- One distinctive reproductive characteristic of common marmosets is the tendency to give birth to twins.
- Both the male and female parents, as well as other group members, participate in caring for and carrying the infants.

Cooperative Parenting:

Cooperative parenting is a crucial aspect of marmoset social groups. Older siblings, in addition to the breeding pair, actively participate in infant care, including carrying and feeding the young.

4.1.4 Social Interactions and Foraging
Cooperative Foraging:

- Marmosets engage in cooperative foraging, where group members work together to locate and access food sources.
- Cooperative foraging enhances the efficiency of resource acquisition and contributes to the overall success of the group.

Territorial Behavior:

Marmoset groups exhibit territorial behavior, marking and defending specific areas within their home range. Territoriality helps ensure access to critical resources, including food and suitable nesting sites.

4.1.5 Significance of Social Groups
Survival and Reproductive Success:

- Social groups contribute to the survival and reproductive success of marmosets by enabling cooperative behaviors and support systems.
- The division of labor within the group enhances the overall well-being of individuals.

Adaptation to Changing Environments:

The social structures of marmoset groups contribute to their adaptability in response to changes in their environments, such as alterations in food availability or the introduction of human-modified landscapes.

Understanding the intricacies of marmoset social groups provides valuable insights into their evolutionary adaptations, cooperative behaviors, and the ecological strategies that contribute to their success in diverse environments. Social interactions play a fundamental role in the lives of these primates, influencing their behavior, communication, and overall well-being.

## Communication explain

4.2 Communication in Marmoset Monkeys

Communication is a fundamental aspect of marmoset monkey behavior, playing a vital role in their social structures, cooperative activities, and overall survival. Marmosets employ a diverse array of vocalizations, body language, and scent-marking to convey information within their social groups. Understanding their communication methods provides insights into their complex social dynamics. Let's explore the various aspects of communication in marmoset monkeys:

## 4.2.1 Vocal Communication
Chirps and Trills:

Marmosets are known for their distinctive vocalizations, which include a variety of chirps, trills, and whistles.
These vocalizations serve multiple purposes, such as signaling alarm, expressing excitement, or coordinating group activities.
Long-Distance Calls:

Marmosets use long-distance calls to communicate with other members of their social group, especially when separated within their home range.
These calls help maintain contact and coordinate movements within the forest canopy.
Food-Related Calls:

Marmosets have specific calls associated with finding food sources. These calls can alert other group members to the presence of a potential food item.
Cooperative foraging is facilitated by these food-related vocalizations.

## 4.2.2 Body Language and Facial Expressions
Tail Movements:

The movement and positioning of the tail are significant components of marmoset body language.

Tail postures may convey emotions, intentions, or signal the readiness to engage in certain behaviors.

Facial Expressions:

Marmosets use facial expressions to convey emotions, including displays of fear, aggression, submission, or contentment.

Expressive features may include eye widening, mouth movements, and changes in the positioning of the ears.

Grooming and Physical Contact:

Social grooming, where individuals groom each other's fur, is not only a hygienic behavior but also a form of communication that reinforces social bonds within the group.

Physical contact, such as huddling or grooming, serves as a means of reassurance and cooperation.

## 4.2.3 Scent-Marking

Territorial Marking:

Marmosets have scent glands on their chest and genital regions that produce scent secretions.

Scent-marking is used to define territories, mark routes, and convey information about individual identity and reproductive status.
Chemical Communication:

Scent-marking plays a crucial role in chemical communication within social groups.
Marmosets can distinguish between the scents of different individuals, aiding in social recognition and coordination.

4.2.4 Context-Specific Communication
Reproductive Signals:

Marmosets use specific vocalizations and body language to signal reproductive readiness or interest.
These signals play a role in coordinating mating behaviors and reproductive activities within the group.
Alarm Calls:

Marmosets have distinct alarm calls that alert the group to potential threats, such as predators or unfamiliar intruders.
Cooperative responses to alarm calls contribute to the group's overall safety.

4.2.5 Social Structure and Communication

Coordinated Behaviors:

Communication is integral to the coordination of group activities, including foraging, territorial defense, and caring for infants.
Cooperative behaviors are facilitated by effective communication within the social structure.
Social Hierarchy:

Communication also plays a role in establishing and maintaining social hierarchies within marmoset groups.
Dominant and subordinate individuals may use specific vocalizations and body language to assert or submit to their social status.
Understanding the nuanced communication methods of marmoset monkeys provides a glimpse into the sophistication of their social interactions and the mechanisms that contribute to the cohesion and success of their social groups. The combination of vocalizations, body language, and scent-marking allows for a rich and complex communication system that enhances their adaptability and survival in their natural environment

**Reproductive Behavior**

4.3 Reproductive Behavior in Marmoset Monkeys

Reproductive behavior in marmoset monkeys is characterized by unique features, including monogamous pair bonding, cooperative breeding, and complex social structures. Marmosets exhibit a range of behaviors related to courtship, mating, gestation, and infant care. Understanding their reproductive strategies provides insights into the evolutionary adaptations that contribute to the success of marmoset social groups.

4.3.1 Monogamous Pair Bonding
Long-Term Pairing:

Marmosets are known for forming long-term monogamous pairs, consisting of one male and one female.
These pairs often remain together throughout their lives, engaging in cooperative behaviors such as foraging, territory defense, and parenting.
Social Bonding:

Monogamous pair bonding contributes to strong social bonds within the family group, enhancing the overall stability and cooperation of the group.
4.3.2 Reproductive Cycle
Estrus and Ovulation:

Female marmosets typically have a defined reproductive cycle, characterized by estrus and ovulation.

Females display behavioral and physiological changes during estrus, signaling their readiness to mate.

Mate Selection:

During estrus, females may engage in behaviors that attract the attention of the male, including vocalizations and body movements.

Monogamous pairs reinforce their bonds through grooming and other social interactions during this period.

### 4.3.3 Courtship and Mating

Courtship Displays:

Marmosets engage in courtship displays, involving vocalizations, facial expressions, and body postures.

Courtship behaviors strengthen the pair bond and facilitate successful mating.

Mating Rituals:

Mating rituals involve specific behaviors, including mutual grooming, vocal exchanges, and physical proximity.

Marmosets often mate within the confines of their established territories.

### 4.3.4 Cooperative Breeding
Twin Births:

One of the distinctive reproductive characteristics of marmosets is the tendency to give birth to twins.
The mother and father, as well as other group members, actively participate in caring for the infants.
Cooperative Infant Care:

Older siblings within the family group play a significant role in infant care, assisting in carrying, grooming, and feeding the newborns.
Cooperative breeding enhances the survival and well-being of the offspring.

### 4.3.5 Parenting Roles
Maternal and Paternal Care:

Both the mother and father are involved in caring for the infants.
Mothers carry and nurse the infants, while fathers actively participate in carrying, grooming, and protecting them.
Sibling Assistance:

Older siblings within the group contribute to infant care, forming a cooperative system that allows for the distribution of parenting responsibilities.

### 4.3.6 Infant Development and Weaning
Gestation and Birth:

The gestation period for marmosets is relatively short, lasting around 4 to 5 months.
After gestation, females give birth to twins.
Weaning Process:

The weaning process is gradual, with infants transitioning to solid food while still receiving care and support from the group.
Older siblings continue to play a role in the socialization and education of the infants.

### 4.3.7 Reproductive Suppression
Inhibition of Reproduction:

In some marmoset species, reproductive suppression occurs among non-breeding individuals within the group.
This suppression prevents breeding competition and contributes to the overall stability of the group.
Role of Hormones:

Hormonal mechanisms, including the inhibition of ovulation in subordinate females, play a role in reproductive suppression.

The dominant breeding pair typically maintains reproductive control within the group.

### 4.3.8 Significance of Reproductive Behavior
Cooperative Strategies:

The cooperative breeding and parenting strategies observed in marmoset monkeys contribute to the survival and success of the social group.

Shared responsibilities among group members enhance the overall well-being of the offspring.

Monogamous Bonding:

Monogamous pair bonding provides stability and support for the reproduction and raising of offspring.

Strong social bonds within the family unit foster cooperation and coordination.

Evolutionary Adaptations:

Marmoset reproductive behaviors reflect evolutionary adaptations that optimize the allocation of resources and increase the chances of offspring survival.

Understanding the intricacies of reproductive behavior in marmoset monkeys offers valuable insights into the complex social structures, cooperative strategies, and adaptive mechanisms that contribute to the success and resilience of these primates in their natural environments.

# CHAPTER FIVE

## Diet and Feeding Habits

**Natural Diet**

5.1 Natural Diet of Marmoset Monkeys

Marmoset monkeys are known for their diverse and adaptable diets, which vary based on their specific habitat and the availability of resources. Their natural diet encompasses a wide range of food items, including fruits, gums, insects, and other small invertebrates. The ability to consume various food sources contributes to their adaptability in different environments. Let's explore the components of the natural diet of marmoset monkeys in greater detail:

5.1.1 Fruits
Variety of Fruits:

Marmosets are frugivores, and fruits form a significant part of their diet.
They consume a variety of fruits, including berries, figs, and other pulpy fruits found in their native habitats.
Role in Nutrition:

Fruits provide essential vitamins, minerals, and sugars that contribute to the overall nutritional balance of the marmoset's diet.

The availability of fruits varies seasonally, influencing the diversity of their diet.

5.1.2 Tree Saps and Gums
Sap Feeding:

Marmosets are known for gouging tree trunks to access sap or gum exudates.

They use their specialized incisor teeth to create holes in tree bark, allowing the sap to flow, which they then lick or consume.

Energy Source:

Tree saps and gums serve as an important energy source, providing carbohydrates and sugars that complement the nutritional content of fruits.

5.1.3 Insects and Invertebrates
Insectivorous Behavior:

Marmosets are also insectivores, actively hunting for insects and other small invertebrates.

They capture insects such as grasshoppers, caterpillars, and spiders using their agile hands and sharp teeth.
Protein Intake:

Insects contribute to the marmoset's protein intake, supporting their dietary requirements for growth, reproduction, and overall health.

5.1.4 Nectar and Flowers
Nectar Consumption:

Marmosets may consume nectar from flowers, especially those with tubular shapes that allow for easy access.
Nectar serves as a liquid energy source, providing sugars for quick energy.
Floral Parts:

Marmosets may also consume other parts of flowers, such as petals or pollen, depending on the species and availability.

5.1.5 Seeds and Plant Parts
Seed Ingestion:

Marmosets may ingest seeds while feeding on fruits, contributing to seed dispersal in their ecosystems.

Seeds that pass through their digestive system may be deposited in new locations, aiding in plant propagation.
Leaves and Plant Exudates:

Some marmoset species may consume leaves or plant exudates in addition to their primary diet of fruits and insects.
The consumption of plant parts provides additional nutrients and may vary based on the species and habitat.

## 5.1.6 Dietary Adaptations
Flexible Diet:

The flexibility of the marmoset's diet allows them to adapt to changes in food availability and environmental conditions.
Their ability to exploit a variety of food sources contributes to their success in diverse habitats.
Geographic Variation:

The specific components of the natural diet can vary among marmoset species and geographic locations.
Adaptations to local food sources contribute to the survival of different populations.

## 5.1.7 Water Intake
Water Requirements:

While marmosets obtain moisture from the foods they consume, they may also drink water directly from natural sources like rivers or rainwater collected in tree hollows.

The need for additional water intake depends on factors such as climate and food moisture content.

Understanding the natural diet of marmoset monkeys highlights their adaptability and resourcefulness in foraging for a wide range of food items. This dietary flexibility contributes to their ecological roles as seed dispersers, insect predators, and key components of the ecosystems they inhabit.

## Captive Diet

5.2 Captive Diet of Marmoset Monkeys

In captivity, providing a balanced and nutritionally appropriate diet is crucial for the health and well-being of marmoset monkeys. The captive diet aims to replicate the nutritional diversity they would experience in their natural habitat. It typically consists of a combination of commercially available primate chow, fresh fruits, vegetables, and other supplements. The goal is to ensure that captive marmosets receive the necessary nutrients for growth, reproduction, and overall health. Let's explore the components and considerations of the captive diet for marmoset monkeys:

## 5.2.1 Commercial Primate Chow
Balanced Nutrition:

Commercial primate chow serves as a base diet that is formulated to provide essential nutrients required for marmoset health.
It typically contains a mix of proteins, carbohydrates, fats, vitamins, and minerals.

Pellet Form:

Primate chow is often provided in pellet form, promoting dental health and reducing selective feeding.
These pellets are carefully formulated to meet the specific dietary needs of marmosets.

## 5.2.2 Fresh Fruits and Vegetables
Variety and Color:

Fresh fruits and vegetables are essential components of the captive marmoset diet, offering a variety of textures, flavors, and colors.
These foods provide additional vitamins, minerals, and dietary fiber.

Seasonal Rotation:

Marmosets benefit from a rotation of fresh produce to mimic seasonal changes in their natural diet.
The rotation helps prevent dietary monotony and ensures a diverse nutrient intake.

### 5.2.3 Insects and Protein Sources
Insectivorous Diet:

Captive marmosets may receive a source of live insects or other protein-rich foods to fulfill their insectivorous dietary needs.
Crickets, mealworms, and other insect species can be offered as part of their diet.

Egg and Lean Meat:

Some captive marmosets may receive cooked eggs or lean meats as additional protein sources.
These foods contribute essential amino acids for muscle development and overall health.

### 5.2.4 Nectar Substitutes
Commercial Nectar Supplements:

Nectar substitutes, often in liquid or powdered form, may be provided as a source of additional sugars.

These supplements mimic the nectar that marmosets may consume in their natural habitat.

5.2.5 Nutritional Supplements
Vitamin and Mineral Supplements:

Depending on the specific dietary requirements and health assessments, marmosets in captivity may receive vitamin and mineral supplements.

These supplements ensure that the diet meets all nutritional needs.

5.2.6 Water Consumption
Access to Fresh Water:

Marmosets in captivity should have access to fresh, clean water at all times.

Hydration is essential for overall health, and water consumption may vary based on factors such as diet and environmental conditions.

5.2.7 Dietary Considerations
Individual Variation:

Individual marmosets may have different dietary preferences, and adjustments may be made based on their specific needs.

Monitoring each individual's response to the diet helps tailor nutritional plans.

Veterinary Oversight:

Regular veterinary check-ups are essential to assess the health and dietary needs of captive marmosets.

Veterinary professionals may provide guidance on dietary adjustments based on the marmoset's age, reproductive status, and overall health.

5.2.8 Behavioral Enrichment

Foraging Opportunities:

Providing opportunities for foraging, such as hiding food items or using puzzle feeders, promotes mental stimulation and satisfies natural foraging behaviors.

Environmental Complexity:

Enclosures with environmental complexity, including structures for climbing and exploring, contribute to the overall well-being of captive marmosets.

The captive diet of marmoset monkeys involves a carefully crafted combination of commercial primate chow, fresh fruits and vegetables, protein sources, and nutritional supplements. Meeting their dietary needs not only supports their physical health but also addresses their behavioral and environmental enrichment requirements. Continuous monitoring, veterinary oversight, and adjustments based on individual needs contribute to the overall health and happiness of marmosets in captivity.

**Feeding Patterns**

5.3 Feeding Patterns of Marmoset Monkeys
The feeding patterns of marmoset monkeys are influenced by various factors, including their natural behaviors, dietary preferences, and the availability of food resources in their habitats. Both in the wild and in captivity, marmosets exhibit distinct feeding behaviors and patterns that contribute to their survival and overall well-being. Let's explore the feeding patterns of marmoset monkeys in detail:

5.3.1 Frequent and Small Meals
Continuous Foraging:

Marmosets are known for their continuous foraging behavior, engaging in frequent but smaller meals throughout the day.

This pattern is consistent with their natural ecology, where they exploit a variety of food sources in their arboreal habitats.

Efficient Energy Utilization:

Frequent feeding allows marmosets to efficiently utilize energy and maintain stable blood sugar levels.
It is adaptive to their small size and high metabolic rates, supporting their active and agile lifestyles.

5.3.2 Arboreal Foraging
Tree Canopy Exploration:

Marmosets are adept climbers and foragers, exploring the tree canopy to find fruits, insects, and tree saps.
Their arboreal foraging behavior involves leaping, climbing, and navigating through the branches in search of food.
Selective Foraging:

Marmosets exhibit selective foraging, choosing specific food items based on availability, preferences, and nutritional content.
The ability to exploit a diverse range of food sources contributes to their dietary flexibility.

### 5.3.3 Gouging for Tree Saps
Gouging Behavior:

Marmosets employ gouging behavior to access tree saps and gums.
They use their specialized incisor teeth to create holes in tree bark, allowing the sap to flow, which then consume.
Energy-Rich Sap Consumption:

Consuming tree saps provides marmosets with an energy-rich food source, contributing to their overall nutritional intake.

### 5.3.4 Social Foraging
Cooperative Foraging:

Marmosets exhibit cooperative foraging behaviors, especially within family groups.
Group members coordinate their activities to locate and access food sources, enhancing the efficiency of resource acquisition.
Food Sharing:

Social foraging may involve food sharing within the group, particularly during activities such as grooming or feeding infants.

Cooperative behaviors contribute to the social bonds within the marmoset family unit.

## 5.3.5 Seasonal Variations
Dietary Adaptations:

Marmosets may exhibit seasonal variations in their feeding patterns, influenced by the availability of certain fruits, insects, or other food resources.
Their ability to adapt to changing seasons contributes to their ecological success.

## 5.3.6 Captive Feeding Patterns

Regular and Varied Meals:

In captivity, marmosets are often provided with regular and varied meals to mimic the diversity of their natural diet.
Captive feeding patterns aim to provide balanced nutrition and address the dietary needs of individual monkeys.

Foraging Opportunities:

Captive environments may include foraging opportunities, such as hiding food items or using puzzle feeders.

Environmental enrichment through foraging activities promotes mental stimulation and engages natural behaviors.

5.3.7 Monitoring and Dietary Adjustments
Veterinary Oversight:

Regular veterinary check-ups are essential to monitor the health and dietary needs of marmosets.
Dietary adjustments may be made based on factors such as age, reproductive status, and overall well-being.
Observation in Captivity:

Caregivers and researchers closely observe marmosets in captivity to understand individual dietary preferences and adjust feeding practices accordingly.
Monitoring behavioral cues helps ensure that each marmoset receives an appropriate and well-balanced diet.
Understanding the feeding patterns of marmoset monkeys provides valuable insights into their ecological adaptations, natural behaviors, and dietary requirements. Whether in the wild or in captivity, their feeding patterns are closely linked to their evolutionary history and play a crucial role in their overall health and survival.

# CHAPTER SIX
Conservation Status

## Threats to Marmoset Populations

6.1 Threats to Marmoset Populations: A Comprehensive Overview

Marmoset populations face numerous threats, both natural and anthropogenic, that can significantly impact their survival and well-being. These threats vary across different species and geographic locations. Understanding these challenges is crucial for the development of effective conservation strategies. Here is a comprehensive overview of the threats to marmoset populations:

6.1.1 Habitat Loss and Fragmentation
Deforestation:

One of the most significant threats to marmoset populations is habitat loss due to deforestation.
Logging, agricultural expansion, and urbanization contribute to the destruction of the marmosets' natural habitats.

Fragmentation:

Habitat fragmentation further exacerbates the problem, isolating populations and limiting their ability to move between fragmented patches of forest.

Fragmentation disrupts ecological processes, such as gene flow and species interactions.

## 6.1.2 Agricultural Expansion

Conversion of Forests:

The expansion of agriculture, including the conversion of forests into farmland, poses a direct threat to marmoset habitats.

Large-scale monoculture practices and agribusiness operations can lead to extensive habitat destruction.

Pesticide Use:

The use of pesticides in agriculture can indirectly harm marmosets by contaminating their food sources and water, leading to poisoning and ecological imbalances.

## 6.1.3 Climate Change

Altered Habitats:

Climate change contributes to alterations in temperature and precipitation patterns, affecting the distribution of plant and animal species, including marmosets.

Shifts in climate can lead to changes in the availability of food resources and disrupt ecological relationships.
Extreme Weather Events:

More frequent and intense extreme weather events, such as hurricanes and droughts, can have detrimental effects on marmoset habitats and populations.

6.1.4 Hunting and Trapping
Bushmeat Trade:

Marmosets are often targeted in the bushmeat trade for their meat or as pets.
Overhunting can lead to population declines, disrupt social structures, and impact ecosystem dynamics.

Capture for the Pet Trade:

The capture of marmosets for the exotic pet trade is a significant threat, contributing to the decline of wild populations.
Captive marmosets may face challenges in adapting to a captive environment.

6.1.5 Disease
Introduction of Diseases:

Human activities, including habitat encroachment, can introduce diseases to marmoset populations.

Diseases such as zoonoses can have devastating effects on marmoset health, leading to population declines.

6.1.6 Invasive Species

Competition and Predation:

The introduction of invasive plant or animal species can disrupt local ecosystems and negatively impact marmosets.

Invasive species may compete for resources or act as predators, further stressing marmoset populations.

6.1.7 Pollution

Water and Air Pollution:

Pollution from industrial activities, agriculture, and urban areas can contaminate water sources and air quality.

Pollutants can affect marmosets directly or indirectly through the contamination of their food and water.

6.1.8 Lack of Legal Protection

Insufficient Conservation Measures:

Inadequate legal protection and enforcement of conservation measures contribute to the vulnerability of marmoset populations.

Weak regulations may fail to address threats effectively or deter illegal activities.

6.1.9 Population Density and Genetic Diversity
Small Population Sizes:

Some marmoset populations may exist in small and isolated groups, making them more susceptible to the effects of environmental changes and genetic issues.
Low genetic diversity within small populations can lead to inbreeding depression and decreased adaptive potential.

6.1.10 Lack of Public Awareness
Educational Deficits:

Lack of public awareness about the ecological importance of marmosets and the threats they face can hinder conservation efforts.
Public education and community involvement are essential for fostering a sense of responsibility for marmoset conservation.
Conservation Efforts and Solutions:
Addressing these threats requires a multi-faceted approach involving habitat protection, sustainable land use practices, legal enforcement, community engagement, and global cooperation. Conservation

efforts should prioritize the preservation of marmoset habitats, the mitigation of climate change impacts, and the establishment of effective protected areas. Additionally, education and outreach programs are crucial for raising awareness about the importance of marmosets and promoting responsible conservation practices.

**Conservation Efforts**

6.2 Conservation Efforts for Marmoset Populations: An In-Depth Overview
Conservation efforts for marmoset populations are critical to mitigating the numerous threats they face. Conservation strategies aim to protect habitats, address human-wildlife conflicts, and involve local communities in sustainable practices. Here's an extensive overview of key conservation efforts for marmoset populations:

6.2.1 Habitat Protection and Restoration
Establishment of Protected Areas:

Designating and effectively managing protected areas is fundamental to safeguarding marmoset habitats.
National parks, reserves, and wildlife sanctuaries help provide secure environments for marmoset populations.

Corridor Creation:

Creating wildlife corridors between fragmented habitats allows marmosets to move freely, promoting genetic diversity and maintaining ecological processes.
Corridors enhance connectivity between isolated populations.
Reforestation Programs:

Implementing reforestation programs helps restore degraded habitats and provides additional resources for marmosets.
Community involvement in tree planting initiatives supports habitat restoration.

6.2.2 Sustainable Land Use Practices
Promoting Agroforestry:

Encouraging agroforestry practices integrates trees into agricultural landscapes, benefiting both farmers and wildlife.
This approach maintains biodiversity while supporting sustainable livelihoods.
Certification Programs:

Certifying sustainable agricultural practices, such as those recognized by organizations like the Rainforest

Alliance, promotes environmentally friendly methods that reduce negative impacts on marmoset habitats.

## 6.2.3 Conservation Education and Awareness
Community Outreach Programs:

Engaging local communities in conservation efforts is crucial. Community outreach programs provide education on the importance of marmosets and sustainable practices.
Workshops, training, and awareness campaigns foster a sense of shared responsibility.
School Programs:

Implementing conservation education programs in schools helps instill environmental awareness from an early age.
Educational initiatives may include curriculum integration, nature walks, and interactive sessions.

## 6.2.4 Research and Monitoring
Population Surveys:

Regular surveys and monitoring programs assess marmoset population sizes, health, and distribution.
Data collected from field studies inform conservation strategies and help track changes over time.

Behavioral Studies:

Understanding marmoset behavior, social structures, and ecological requirements through behavioral studies aids in designing targeted conservation interventions.

## 6.2.5 Anti-Poaching Measures
Enforcement of Laws:

Strengthening and enforcing laws against hunting and the illegal pet trade is essential.
Cooperation with law enforcement agencies and the judiciary helps prosecute offenders.
Community-Based Monitoring:

Involving local communities in monitoring and reporting illegal activities promotes a sense of ownership and contributes to anti-poaching efforts.

## 6.2.6 Disease Management
Health Surveillance:

Implementing health surveillance programs for marmoset populations helps detect and manage disease outbreaks.
Quarantine measures may be employed to prevent the spread of diseases.

6.2.7 Conservation Partnerships
Collaboration with NGOs and Governments:

Collaborations with non-governmental organizations (NGOs), governmental agencies, and international conservation bodies strengthen conservation initiatives.
Joint efforts can leverage resources, expertise, and global support.
Partnerships with Indigenous Communities:

Collaborating with indigenous communities, who often have traditional knowledge and a strong connection to local ecosystems, fosters sustainable conservation practices.

6.2.8 Climate Change Adaptation
Climate-Resilient Conservation:

Integrating climate change adaptation strategies into conservation planning helps marmoset populations cope with changing environmental conditions.
This may involve creating climate-resilient habitats and promoting sustainable land use practices.

6.2.9 Genetic Management
Genetic Monitoring:

Genetic monitoring programs help assess the genetic diversity of marmoset populations.

Strategies for maintaining genetic health, such as translocations between populations, can be implemented if necessary.

6.2.10 Sustainable Tourism Practices
Eco-Tourism Initiatives:

Promoting eco-tourism initiatives that prioritize wildlife conservation and responsible visitor behavior generates economic benefits for local communities.

Sustainable tourism practices help reduce negative impacts on marmoset habitats.

Conservation efforts for marmoset populations require a holistic and collaborative approach that addresses habitat protection, sustainable land use practices, community engagement, and the mitigation of various threats. The involvement of local communities, governments, NGOs, and international organizations is essential for the success of conservation initiatives. By implementing comprehensive strategies, we can work towards ensuring the long-term survival and well-being of marmoset populations and their ecosystems.

**Importance of Conservation**

6.3 Importance of Conservation for Marmoset Populations: A Comprehensive Overview

Conservation efforts for marmoset populations are of paramount importance due to the crucial roles these primates play in maintaining ecological balance, biodiversity, and the overall health of ecosystems. The significance of marmoset conservation extends beyond individual species to the broader landscapes they inhabit. Here's a comprehensive overview of the importance of conservation for marmoset populations:

6.3.1 Biodiversity Preservation
Ecosystem Engineers:

Marmosets, as part of their ecosystems, contribute to biodiversity preservation by serving as pollinators, seed dispersers, and contributors to nutrient cycling.
Their ecological roles are essential for maintaining the health and diversity of plant and animal communities.
Keystone Species:

Marmosets can be considered keystone species, meaning their presence has a disproportionate impact on the structure and function of their ecosystems.

Changes in marmoset populations can cascade through the food web, affecting various species.

### 6.3.2 Ecological Balance
Predator-Prey Dynamics:

Marmosets contribute to the regulation of insect populations through their insectivorous behaviors, helping control pest species.
Their presence contributes to a balanced predator-prey dynamic within their ecosystems.
Seed Dispersal:

Marmosets play a crucial role in seed dispersal by consuming fruits and then excreting seeds in different locations.
This behavior aids in the regeneration of plant species and helps maintain healthy forests.

### 6.3.3 Genetic Diversity
Genetic Resilience:

Maintaining genetically diverse marmoset populations is essential for their long-term resilience to environmental changes.

Genetic diversity enhances their ability to adapt to evolving threats, including diseases and habitat alterations.
Population Health:

Genetic diversity is linked to overall population health, reducing the risk of inbreeding and associated genetic disorders.
Healthy populations contribute to the stability and adaptability of the species.

6.3.4 Scientific Research
Biological Insights:

Marmosets serve as valuable subjects for scientific research, providing insights into primate behavior, cognition, and social structures.
Studying marmoset populations contributes to our understanding of evolutionary processes and ecological interactions.
Medical Research Models:

Marmosets are used as models in medical research due to their physiological similarities to humans.
Conservation efforts ensure the availability of healthy and genetically diverse populations for research purposes.

### 6.3.5 Cultural and Aesthetic Value
Cultural Significance:

Marmosets hold cultural significance for many indigenous communities, forming part of local folklore, traditions, and spiritual beliefs.
Conservation efforts respect and preserve the cultural importance of marmosets to these communities.
Aesthetic Enjoyment:

Marmosets contribute to the aesthetic value of natural environments, offering opportunities for eco-tourism and providing enjoyment for those appreciating wildlife in its natural habitat.

### 6.3.6 Ethical Considerations
Moral Responsibility:

The ethical dimension of conservation recognizes the intrinsic value of all species and ecosystems.
Protecting marmoset populations reflects a moral responsibility to safeguard biodiversity for future generations.
Interconnectedness of Life:

Conservation acknowledges the interconnectedness of all life forms and recognizes that the well-being of marmosets is intertwined with the health of ecosystems and the planet as a whole.

6.3.7 Climate Change Resilience

Adaptation Strategies:

Conserving marmoset populations contributes to overall ecosystem resilience against the impacts of climate change.

Healthy ecosystems are better equipped to adapt to changing environmental conditions.

Carbon Sequestration:

Forests inhabited by marmosets play a role in carbon sequestration, helping mitigate climate change by capturing and storing carbon dioxide.

The importance of marmoset conservation goes beyond the preservation of a single species; it encompasses the well-being of ecosystems, the sustenance of biodiversity, and the ethical responsibility to protect the diversity of life on Earth. Conservation efforts contribute to ecological balance, genetic diversity, scientific knowledge, and the cultural and aesthetic values associated with these fascinating primates. By prioritizing marmoset conservation, we contribute to the broader mission of preserving the intricate web of life that sustains our planet.

# CHAPTER SEVEN

## Interaction with Humans

**Marmosets as Pets**

7.1 Marmosets as Pets: Considerations and Ethical Concerns
Keeping marmosets as pets has gained popularity in certain regions, driven by their small size, charming appearance, and perceived intelligence. However, the trend of keeping marmosets as pets raises numerous ethical, legal, and welfare concerns. Here's an extensive exploration of the considerations and ethical concerns associated with keeping marmosets as pets:

7.1.1 Legal Considerations
Permits and Regulations:

Many countries and regions have strict regulations regarding the ownership of primates, including marmosets, as pets.
Permits and licenses may be required to ensure compliance with wildlife protection laws.

Illegal Trade:

The demand for marmosets as pets can contribute to illegal wildlife trade.
Purchasing marmosets from unregulated sources may perpetuate the exploitation and trafficking of these primates.

7.1.2 Ethical Concerns
Wildlife Exploitation:

Capturing marmosets from the wild for the pet trade can have devastating effects on wild populations.
Removal of individuals disrupts social structures and can lead to population declines.
Domestication Challenges:

Marmosets are wild animals with complex social structures and specialized dietary and environmental needs.
Domestication challenges include difficulties meeting their physical, social, and psychological requirements in a home setting.

7.1.3 Welfare Issues
Social Isolation:

Marmosets are highly social animals that thrive in family groups.
Keeping them as solitary pets can lead to social isolation, affecting their mental and emotional well-being.

Health Concerns:

Marmosets have specific dietary requirements and are prone to health issues in captivity, including obesity and dental problems.
Maintaining their health necessitates expert knowledge and access to appropriate veterinary care.

7.1.4 Zoonotic Risks

Disease Transmission:

Primates, including marmosets, can carry zoonotic diseases that can be transmitted to humans.
Close contact with pet marmosets may pose health risks to owners, especially if proper hygiene practices are not observed.

7.1.5 Behavioral Challenges

Aggressive Behavior:

Marmosets may exhibit aggressive behaviors, especially as they reach sexual maturity.

Their natural behaviors, including vocalizations and scent marking, may pose challenges in a domestic setting.
Longevity and Commitment:

Marmosets have a relatively long lifespan in captivity, often living over 10 years.
Owners must commit to providing lifelong care, including meeting their evolving needs as they age.

7.1.6 Conservation Perspective
Impact on Wild Populations:

The pet trade can contribute to the depletion of wild marmoset populations.
Conservation efforts are compromised when individuals are taken from the wild for the pet trade.
Educational Alternatives:

Advocates argue that promoting education and public awareness about marmosets in their natural habitats is a more ethical way to appreciate and support their conservation.

7.1.7 Alternatives to Pet Ownership
Supporting Conservation Organizations:

Individuals passionate about marmosets can contribute to their conservation by supporting reputable organizations engaged in habitat protection, research, and education. Responsible Tourism:

Eco-tourism initiatives that prioritize responsible wildlife viewing allow individuals to appreciate marmosets in their natural environments without compromising their welfare.

While the appeal of having a marmoset as a pet is understandable, the ethical concerns, legal restrictions, and welfare considerations associated with such practices are substantial. Conservationists and animal welfare advocates emphasize the importance of appreciating marmosets in their natural habitats and supporting initiatives that contribute positively to their conservation. Responsible alternatives, such as promoting awareness and supporting conservation organizations, can foster a greater understanding and appreciation of these remarkable primates without compromising their well-being and the health of wild populations.

## Ethical Considerations

7.2 Ethical Considerations of Keeping Marmosets as Pets
The ethical considerations surrounding the keeping of marmosets as pets are multifaceted, encompassing issues

related to the welfare of the animals, conservation implications, and broader ethical concerns associated with the pet trade. Here's an in-depth exploration of the ethical considerations:

7.2.1 Welfare of Marmosets
Complex Social Structures:

Marmosets are highly social animals that live in family groups in the wild.
Keeping them as solitary pets denies them the opportunity to engage in natural social behaviors, leading to stress and behavioral issues.
Specialized Dietary and Environmental Needs:

Marmosets have specific dietary requirements and need access to a complex environment that includes climbing structures and enrichment.
Captive environments often struggle to meet these needs, impacting the physical and mental well-being of the animals.

Health Concerns:

Captive marmosets may face health issues such as obesity, dental problems, and stress-related conditions.

Providing appropriate veterinary care and meeting their nutritional and medical needs can be challenging in a home setting.

## 7.2.2 Conservation Implications
Impact on Wild Populations:

The demand for marmosets as pets can contribute to the illegal wildlife trade, leading to the capture and removal of individuals from their natural habitats.
This practice poses a direct threat to wild populations, disrupting social structures and reducing genetic diversity.

Conservation Disconnect:

Keeping marmosets as pets may create a disconnect between individuals and the broader conservation challenges facing these species in the wild.
Conservation efforts are more effective when directed toward preserving natural habitats and addressing the root causes of population decline.

## 7.2.3 Legal and Regulatory Compliance
Permit Requirements:

Many countries and regions have strict regulations regarding the ownership of primates, and obtaining the necessary permits is essential.

Ignoring legal requirements contributes to the illegal pet trade and may lead to negative consequences for both the animals and owners.

Illegal Trade:

Purchasing marmosets from unregulated sources or engaging in the illegal pet trade supports wildlife exploitation and perpetuates the market for wild-caught animals.

## 7.2.4 Zoonotic Risks
Disease Transmission:

Primates, including marmosets, can carry zoonotic diseases that may be transmitted to humans.

Close contact with pet marmosets without proper hygiene practices poses health risks for owners.

## 7.2.5 Behavioral Challenges
Aggressive Behavior:

Marmosets, particularly as they reach sexual maturity, may exhibit aggressive behaviors.

Owners may not be adequately prepared to handle these behaviors, leading to welfare concerns for both the animals and humans.

Incompatibility with Domestic Life:

Marmosets have natural behaviors such as vocalizations, scent marking, and climbing that may be incompatible with a typical domestic setting.
This can lead to challenges in providing an environment that meets their behavioral needs.

## 7.2.6 Long-Term Commitment
Lifespan and Changing Needs:

Marmosets have a relatively long lifespan in captivity, often living over a decade.
Owners must commit to providing lifelong care, including adapting to the changing needs of the animals as they age.

## 7.2.7 Educational Alternatives
Promoting Responsible Conservation:

Ethical considerations suggest that promoting responsible wildlife conservation practices, habitat protection, and public awareness are more sustainable ways to appreciate and support marmosets.

Conservation Education:

Education programs that focus on the natural behaviors, ecology, and conservation status of marmosets can foster a deeper understanding and appreciation for these species without compromising their welfare.

The ethical considerations surrounding the keeping of marmosets as pets underscore the importance of prioritizing the welfare of these animals, adhering to legal and regulatory standards, and promoting responsible conservation practices. Balancing the desire for a pet with the ethical implications of wildlife ownership requires a thorough understanding of the complex needs of marmosets and a commitment to their well-being both in captivity and in the wild. Responsible alternatives, such as supporting conservation efforts and engaging in educational initiatives, contribute positively to the ethical treatment of marmosets.

## Human-Marmoset Conflict

## 7.3 Human-Marmoset Conflict: Causes, Impacts, and Mitigation Strategies

Human-marmoset conflict arises when interactions between humans and marmosets result in negative consequences for both parties. This conflict can have various causes and impacts, ranging from habitat encroachment to health risks. Implementing effective mitigation strategies is crucial to ensure the coexistence of humans and marmosets. Here's an in-depth exploration of human-marmoset conflict:

### 7.3.1 Causes of Human-Marmoset Conflict

Habitat Loss and Fragmentation:

The expansion of human settlements often leads to habitat loss and fragmentation, forcing marmosets to adapt to urban or peri-urban environments.
Reduced access to natural habitats can result in increased contact with human activities.

Feeding on Human Resources:

Marmosets may turn to human resources for food, such as raiding crops, fruit trees, or scavenging in residential areas.
This behavior can lead to conflicts with farmers or residents.

Urbanization and Infrastructure Development:

Urban development and infrastructure projects can alter marmoset habitats, bringing them into closer proximity to human activities.
Increased encounters with humans and traffic pose risks to marmosets.

Pet Trade and Ownership:

The illegal pet trade may lead to marmosets being kept as pets in residential areas.
Escaped or released pets may form feral populations, causing conflicts with humans.

7.3.2 Impacts of Human-Marmoset Conflict
Health Risks:

Close contact between humans and marmosets can pose health risks, as primates can carry zoonotic diseases that may be transmitted to humans.
Bites or scratches during conflicts can also result in injuries and infections.

Crop Damage:

Marmosets foraging for food in agricultural areas may cause damage to crops.
This can lead to economic losses for farmers and exacerbate conflict.

Traffic Accidents:

Marmosets adapting to urban environments may encounter roads, leading to the risk of traffic accidents.
Collisions with vehicles can result in injuries or fatalities for the marmosets and pose safety concerns for drivers.
Stress on Marmoset Populations:

Frequent conflicts with humans can cause stress to marmoset populations, impacting their health, behavior, and reproductive success.
Continuous disturbances may disrupt natural behaviors and social structures.

7.3.3 Mitigation Strategies
Habitat Protection and Restoration:

Protecting natural habitats and restoring degraded areas can minimize habitat loss and fragmentation, reducing the likelihood of conflicts.
Establishing wildlife corridors helps marmosets move safely between fragmented habitats.

Educational Programs:

Implementing educational programs for local communities can increase awareness about marmoset behavior, ecological roles, and the importance of coexistence.
Educated communities are more likely to adopt practices that minimize conflicts.
Responsible Waste Management:

Proper waste management reduces the availability of human food sources for marmosets in urban areas.
Securing trash bins and disposing of waste appropriately can discourage foraging behavior.
Crop Protection Measures:

Implementing crop protection measures, such as fencing or deterrents, can reduce crop damage caused by marmosets.
Farmers can also explore alternative crops that are less attractive to marmosets.

Wildlife-friendly Urban Planning:

Urban planning that considers the needs of wildlife, including green spaces and wildlife corridors, can minimize conflicts in urban environments.

Designing neighborhoods with green infrastructure benefits both humans and marmosets.

Law Enforcement and Regulation:

Strict enforcement of laws against illegal pet trade and ownership helps reduce the likelihood of marmosets being kept as pets.

Implementing and enforcing regulations regarding wildlife protection contribute to conflict prevention.

Health Monitoring and Research:

Regular health monitoring of marmoset populations can help identify and address disease risks.

Research on marmoset behavior and ecology informs the development of targeted mitigation strategies.

Community Engagement:

Engaging local communities in conservation and coexistence efforts fosters a sense of shared responsibility.

Involving residents in decision-making processes promotes sustainable practices.

Human-marmoset conflict is a complex issue influenced by habitat changes, human activities, and the behavior of both parties. Mitigation strategies should focus on addressing the root causes, promoting coexistence, and involving local communities in conservation efforts. Balancing the needs of humans and marmosets requires a collaborative and multidisciplinary approach to ensure the well-being of both species and the preservation of ecosystems.

# CHAPTER EIGHT

## Research and Scientific Significance

**Contributions to Scientific Understanding**

8.1 Contributions of Marmosets to Scientific Understanding: An Extensive Overview

Marmosets, particularly common marmosets (Callithrix jacchus) and other New World monkeys, have made significant contributions to scientific understanding across various fields. Their unique biological features, social structures, and behavioral characteristics make them valuable subjects for research. Here's an extensive exploration of the contributions of marmosets to scientific understanding:

8.1.1 Biomedical and Medical Research
Model Organisms:

Marmosets are used as non-human primate models in biomedical and medical research.

Their physiological similarities to humans, including reproductive systems and brain structures, make them valuable for studying various diseases and disorders.

Neuroscience and Brain Function:

Marmosets are particularly useful in neuroscience research due to their relatively small brain size and well-defined cortical areas.
Studies on marmosets have provided insights into brain function, cognition, and neural pathways, contributing to our understanding of neurological disorders.

Reproductive Biology:

Marmosets have unique reproductive characteristics, including twinning and cooperative breeding.
Research on marmoset reproductive biology has implications for understanding human reproduction and fertility.

8.1.2 Genetics and Genomic Research
Genetic Diversity:

Studying the genetic diversity of marmoset populations contributes to our understanding of primate evolution.

The relatively high genetic diversity within marmoset species provides insights into their adaptive capabilities.

Genetic Engineering:

Advances in genetic engineering technologies, including transgenic and knockout marmosets, enable researchers to investigate the role of specific genes in health and disease.
These genetic tools have implications for developing therapies and understanding genetic contributions to human conditions.

8.1.3 Social Behavior and Communication Studies
Cooperative Breeding and Parental Care:

Marmosets exhibit cooperative breeding, where multiple individuals contribute to the care of offspring.
Studies on marmoset parental care provide insights into the evolution of social structures and cooperation in primates.

Communication Systems:

Marmosets have complex communication systems involving vocalizations, facial expressions, and body language.

Research on marmoset communication contributes to the understanding of primate communication and social dynamics.

## 8.1.4 Aging and Age-Related Diseases
Aging Models:

Marmosets serve as models for studying aging and age-related diseases due to their relatively short lifespan.
Research on marmosets contributes to understanding the physiological changes associated with aging.
Neurodegenerative Diseases:

Studies on marmosets have provided insights into neurodegenerative diseases, such as Parkinson's disease.
Marmoset models are valuable for testing potential treatments and understanding the underlying mechanisms of these diseases.

## 8.1.5 Cognitive and Behavioral Research
Learning and Memory:

Marmosets have been used to study learning and memory processes, contributing to our understanding of cognition in primates.
Research on marmosets has implications for cognitive neuroscience and psychology.

Social Cognition:

Marmosets display advanced social cognition, including cooperation and understanding of social relationships. Studies on social cognition in marmosets provide insights into the evolution of social intelligence in primates.

8.1.6 Pharmacological Research
Drug Testing:

Marmosets are utilized in pharmacological research for testing the safety and efficacy of drugs. Their physiological similarities to humans make them valuable models for assessing drug responses and potential side effects.

Preclinical Trials:

Marmoset models play a role in preclinical trials for developing therapeutic interventions in areas such as neuroscience, infectious diseases, and reproductive health.

8.1.7 Conservation Biology and Ecology
Habitat Studies:

Research on marmoset habitats contributes to conservation biology, providing information on ecological requirements and the impact of habitat changes.

Understanding the ecology of marmosets aids in conservation efforts to protect natural habitats.

Population Dynamics:

Studies on marmoset population dynamics, including behavior, reproduction, and social structures, contribute to conservation strategies and management plans.

Marmosets have become invaluable contributors to scientific understanding across a broad spectrum of disciplines. Their unique biological features, genetic diversity, and complex social behaviors make them excellent models for biomedical, genetic, cognitive, and ecological research. Insights gained from studies on marmosets have direct implications for human health, neuroscience, conservation biology, and our broader understanding of primate evolution and behavior. The ethical and responsible use of marmosets in research continues to enhance our knowledge and contribute to advancements in various scientific fields.

**Medical Research and Applications**

8.2 Medical Research and Applications of Marmosets:

An In-Depth Exploration

Marmosets, particularly the common marmoset (Callithrix jacchus), have become instrumental in advancing medical research and applications. Their biological similarities to humans, relatively small size, and unique reproductive features make them valuable models for studying various diseases and conditions. Here's an extensive exploration of the medical research and applications of marmosets:

8.2.1 Biomedical Models and Comparative Medicine
Physiological Similarities to Humans:

Marmosets share numerous physiological similarities with humans, including reproductive systems, immune responses, and brain structures.
These similarities make marmosets valuable models for studying human health and diseases.

Comparative Medicine Advantages:

Marmosets are advantageous in comparative medicine, allowing researchers to bridge the gap between rodent models and larger non-human primates.

Their small size facilitates housing, breeding, and experimentation.

## 8.2.2 Neuroscience and Brain Disorders
Brain Imaging Studies:

Marmosets are used in neuroscience research, particularly in brain imaging studies, to understand neural circuits and functions.
Insights gained from marmoset models contribute to our understanding of neurodevelopmental disorders and psychiatric conditions.

Neurodegenerative Diseases:

Marmosets serve as models for studying neurodegenerative diseases such as Parkinson's and Alzheimer's.
Research on marmosets provides insights into disease mechanisms, potential treatments, and the development of therapeutic interventions.

## 8.2.3 Reproductive Biology and Fertility Research
Reproductive Characteristics:

Marmosets exhibit unique reproductive characteristics, including twinning and cooperative breeding.

Studies on marmoset reproduction contribute to our understanding of fertility, reproductive health, and developmental biology.

Assisted Reproductive Technologies:

Marmosets are used in the development and refinement of assisted reproductive technologies (ART), including in vitro fertilization (IVF) and embryo transfer.
Insights gained from marmoset models have implications for human reproductive medicine.

### 8.2.4 Genetic Engineering and Transgenic Models
Transgenic Marmosets:

Advances in genetic engineering have allowed the development of transgenic marmoset models.
These models are crucial for studying the role of specific genes in health, disease, and therapeutic interventions.
Gene Editing Technologies:

The application of gene editing technologies, such as CRISPR-Cas9, in marmosets enables the targeted modification of genes for functional studies and disease modeling.

### 8.2.5 Infectious Disease Research
Viral Infection Models:

Marmosets are used in infectious disease research, serving as models for studying viral infections such as Zika virus, dengue virus, and HIV.

Insights from marmoset models contribute to understanding host-pathogen interactions and developing antiviral strategies.

Vaccine Development:

Marmosets play a role in preclinical trials for vaccine development, testing vaccine candidates for safety and efficacy.

Their immune responses provide valuable data for designing effective vaccines.

8.2.6 Pharmacological Studies and Drug Testing
Drug Metabolism and Responses:

Marmosets are used in pharmacological studies to investigate drug metabolism, pharmacokinetics, and responses.

Their physiological similarities to humans make them reliable models for predicting drug effects.

Preclinical Trials:

Marmosets serve as preclinical models in drug development, helping assess the safety and efficacy of potential therapeutics before human trials.
This approach aids in minimizing risks and optimizing drug formulations.

## 8.2.7 Aging and Age-Related Diseases
Aging Models:

Marmosets, with their relatively short lifespan, are employed as models for studying aging and age-related diseases.
Research on marmosets contributes to understanding the physiological changes associated with aging.

Neurological Disorders in Aging:

Marmosets provide insights into age-related neurological disorders, including cognitive decline and neurodegeneration.
Studies contribute to identifying potential interventions to promote healthy aging.

The use of marmosets in medical research has advanced our understanding of various health-related aspects, from neurobiology and reproductive health to infectious diseases and drug development. The unique features of marmosets, combined with their genetic proximity to humans, make them invaluable models for translational research. The ethical and responsible application of marmosets in medical studies continues to yield insights that have direct implications for improving human health, developing therapeutic interventions, and advancing our knowledge of complex diseases.

## Challenges in Marmoset Research

8.3 Challenges in Marmoset Research: Navigating Complexities for Scientific Advancement
While marmosets have proven to be valuable models in various scientific disciplines, their use in research comes with its own set of challenges. These challenges span ethical considerations, logistical issues, and limitations in experimental techniques. Here's an extensive exploration of the challenges associated with marmoset research:

8.3.1 Ethical Considerations
Animal Welfare Concerns:

Marmosets, like all animals used in research, raise ethical concerns related to their well-being.
Balancing the potential benefits of research with the welfare of individual animals is a constant challenge.

Long-Term Care Requirements:

Marmosets have relatively long lifespans in captivity, and ensuring their well-being throughout their lives requires significant resources and commitment.
Ethical considerations include providing appropriate social structures, environmental enrichment, and veterinary care.

8.3.2 Genetic and Biological Variability
Genetic Differences Across Populations:

Marmoset populations may exhibit genetic variability, and studies may involve animals with different genetic backgrounds.
This variability can introduce challenges in drawing generalizable conclusions, and researchers must carefully consider genetic diversity in study design.

Limited Genetic Manipulation Techniques:

While genetic engineering techniques have advanced, the ability to manipulate marmoset genomes is not as refined as with other model organisms.
Challenges in creating specific genetic modifications may limit the scope of studies.

### 8.3.3 Technical and Experimental Constraints
Size Limitations:

The small size of marmosets presents challenges in certain experimental techniques and imaging modalities. Performing surgeries or collecting specific tissue samples may be more challenging compared to larger primate models.

Limited Behavioral Repertoire:

Marmosets, while displaying complex social behaviors, may have a more limited behavioral repertoire compared to larger primates.
Designing experiments that capture a range of cognitive and social behaviors can be challenging.

### 8.3.4 Reproductive Characteristics
Twinning and Cooperative Breeding:

While unique and valuable for reproductive studies, the natural twinning and cooperative breeding characteristics of marmosets can complicate experimental designs.
Controlling variables related to reproduction may require specialized protocols.

Slow Reproductive Maturation:

Marmosets have a relatively slow reproductive maturation compared to rodents, which may extend the timeline for longitudinal studies.
Planning and executing long-term studies require careful consideration of the extended developmental periods.

## 8.3.5 High Costs and Resource Intensity
Housing and Care Costs:

Providing appropriate housing, veterinary care, and enrichment for marmosets can be resource-intensive.
High costs associated with long-term care and maintenance can be a limiting factor for some research programs.

Availability of Specialized Facilities:

Specialized facilities equipped for the housing and care of marmosets are essential for ethical and responsible research.
Limited access to such facilities can restrict the scalability of marmoset research programs.

## 8.3.6 Regulatory Compliance and Permitting
Stringent Regulatory Requirements:

Marmoset research often involves navigating stringent regulatory requirements and obtaining permits.
Researchers must adhere to ethical guidelines and legal frameworks, adding complexity to study initiation and execution.

Import and Export Challenges:

International collaborations may face challenges related to the import and export of marmosets due to regulatory restrictions.
Coordinating cross-border research efforts requires careful planning and compliance with relevant regulations.

Marmoset research, while immensely valuable, is not without its challenges. Addressing ethical considerations, managing genetic and biological variability, navigating technical constraints, and ensuring regulatory compliance are essential aspects of responsible marmoset research. As the scientific community continues to refine techniques, enhance ethical standards, and develop collaborative frameworks, the potential for marmosets to contribute to groundbreaking discoveries remains promising. Balancing the benefits of research with the welfare of individual animals is a shared responsibility among researchers, institutions, and ethical oversight bodies.

# CHAPTER NINE

## Popular Culture

**Marmosets in Art and Literature**

9.1 Marmosets in Art and Literature: A Creative Exploration

Marmosets, with their distinctive appearance and engaging behaviors, have captivated the artistic and literary imagination throughout history. From paintings and illustrations to literature and popular culture, marmosets have been portrayed in various ways, symbolizing different themes and eliciting diverse emotions. Here's an extensive exploration of the presence of marmosets in art and literature:

9.1.1 Artistic Depictions
Exotic and Playful Subjects:

Marmosets, with their small size, colorful fur, and expressive faces, are often depicted in art as exotic and playful subjects.

Artists use their unique features to create visually appealing and whimsical compositions.

Naturalistic Illustrations:

Naturalists and scientific illustrators have depicted marmosets with a focus on accuracy and detail, showcasing their biological characteristics.
Illustrations in natural history books and scientific journals contribute to the workation of marmoset species.

9.1.2 Symbolism in Art
Allegorical Representations:

Marmosets are sometimes used allegorically in art to symbolize specific virtues or vices.
Their social behaviors, such as cooperative breeding, may be interpreted symbolically in the context of human relationships.

Cultural Symbolism:

In some cultures, marmosets may hold symbolic significance.
Artists may incorporate cultural symbolism, mythology, or folklore related to marmosets in their works.

9.1.3 Literary References
In Fiction and Poetry:

Marmosets make appearances in fiction and poetry, often as exotic and intriguing creatures.
Writers use their characteristics to evoke themes of curiosity, playfulness, or the mysterious.
Metaphorical Use:

Marmosets may be employed metaphorically in literature to convey ideas or emotions.
Their social behaviors and interactions may serve as metaphors for human relationships or societal dynamics.
9.1.4 Children's Literature and Media

Adventurous Characters:

In children's literature and media, marmosets are sometimes portrayed as adventurous characters.
Their small size and playful nature make them relatable and endearing to young audiences.

Educational Content:

Marmosets may be featured in educational children's books and programs to introduce young readers to the diversity of the animal kingdom.

Such representations contribute to early awareness of wildlife.

### 9.1.5 Cultural Representations
In Indigenous Art:

In some indigenous art traditions, marmosets may be featured as symbols or characters with cultural significance.
The portrayal of marmosets in indigenous art reflects the relationship between communities and local fauna.

In Modern and Contemporary Art:

Contemporary artists often use marmosets as subjects in a variety of media, exploring themes ranging from biodiversity to the impact of human activities on wildlife. Marmosets may be integrated into installations, sculptures, and multimedia artworks.

### 9.1.6 Art Conservation and Advocacy
Wildlife Conservation Campaigns:

Artists may contribute to wildlife conservation efforts by using marmosets as subjects in campaigns or artworks raising awareness about the conservation status of these primates.

Art can be a powerful tool in advocating for the protection of marmoset habitats.

Ethical Considerations in Art:

Artists may also address ethical considerations related to the portrayal of wildlife, including marmosets, in their work.
The impact of art on public perceptions and conservation attitudes is recognized, prompting discussions on responsible artistic representation.

Marmosets, with their charm and uniqueness, have found a place in various forms of creative expression. Artists and writers draw inspiration from the intriguing qualities of marmosets to convey aesthetic, symbolic, and cultural meanings. Whether depicted in naturalistic illustrations, used metaphorically in literature, or featured in modern art installations, marmosets continue to be a source of inspiration and contemplation, contributing to the rich tapestry of human creativity. As awareness of wildlife conservation grows, the portrayal of marmosets in art and literature also serves as a reflection of society's evolving relationship with the natural world and the ethical considerations associated with representing wildlife.

## Marmosets in Media and Entertainment

9.2 Marmosets in Media and Entertainment: From Film to Mascots

Marmosets, with their charming appearance and playful demeanor, have made notable appearances in various forms of media and entertainment. From feature films and television shows to advertisements and even as mascots, these primates have captured the audience's attention and become memorable figures in popular culture. Here's an extensive exploration of the presence of marmosets in media and entertainment:

9.2.1 Film and Television
Feature Films:

Marmosets have been featured in both live-action and animated films, often serving as adorable and comedic characters.
Their small size and expressive faces make them well-suited for roles in family-friendly films.

Television Shows:

Marmosets may make appearances in nature workaries or wildlife programs, showcasing their natural behaviors and habitats.

In fictional television shows, they can be portrayed as exotic pets or characters in comedic or adventure settings.

## 9.2.2 Advertising and Commercials
Product Advertisements:

Marmosets are occasionally used in advertising to promote products, particularly those related to nature, wildlife, or children's products.
Their cuteness and playful demeanor make them attention-grabbing in commercials.
Brand Mascots:

Some brands use marmosets as mascots to create a distinctive and memorable brand image.
The association with these small, endearing primates can evoke positive emotions and contribute to brand recognition.

## 9.2.3 Social Media and Viral Content
Online Platforms:

Marmosets gain popularity on social media platforms where users share videos and images showcasing their antics and interactions.

Viral content featuring marmosets often garners widespread attention and engagement.

Digital Influencers:

Some marmosets have become digital influencers, with dedicated social media accounts managed by their owners.
Followers are entertained by regular updates on the daily lives and activities of these charismatic primates.

9.2.4 Video Games and Virtual Worlds
Video Game Characters:

Marmosets can be featured as characters in video games, either as part of the game's narrative or as virtual pets.
Their animated counterparts often reflect their real-life characteristics, bringing a touch of nature to virtual worlds.
Virtual Reality Experiences:

Virtual reality applications may include experiences featuring marmosets in their natural habitats, offering users an immersive encounter with these animals.

9.2.5 Educational Programs
Wildlife workaries:

Marmosets are showcased in wildlife workaries, offering viewers insights into their behaviors, habitats, and the challenges they face.

Educational programs may highlight conservation efforts aimed at protecting marmoset populations.

Children's Educational Content:

Marmosets are incorporated into children's educational content, contributing to early awareness about biodiversity and the importance of wildlife conservation.

9.2.6 Cultural References in Entertainment

Literary Adaptations:

Marmosets in literature may find their way into film or television adaptations, bringing beloved characters to life on screen.

The visual representation of marmosets often draws inspiration from their literary descriptions.

Animated Series:

Marmosets may be featured in animated series, appealing to a broad audience with their entertaining antics and relatable personalities.

Marmosets have carved a niche for themselves in the realm of media and entertainment, charming audiences across various platforms. From their roles in films and television to their appearances in advertising, social media, and virtual environments, these primates contribute to the diverse tapestry of popular culture. While their endearing qualities make them popular figures in entertainment, it's essential to consider the responsible portrayal of marmosets, acknowledging their status as wildlife and promoting awareness about the importance of ethical treatment and conservation efforts.

# CHAPTER TEN

## Conclusion

10.1 Summary of Key Points

10.1 Summary of Key Points
Marmoset Overview:
Scientific Classification:

Marmosets belong to the family Callitrichidae and are New World monkeys, with common species like Callithrix jacchus.
They are characterized by small size, claw-like nails, and a dental formula that includes specialized incisors.
Habitat and Distribution:

Marmosets inhabit diverse environments, from tropical rainforests to savannas, primarily in South America.
Species-specific adaptations allow them to thrive in varied ecological niches.
Biological Characteristics:
Size and Weight:

Marmosets are small primates, with an average body length of 5-12 inches and weighing 3-16 ounces.

Fur and Coloration:

Fur coloration varies among species, with patterns of black, brown, gray, and white.
Some species have manes or tufts on their ears.

Tail Characteristics:

Marmosets have long tails, often non-prehensile, which aid in balance and locomotion.
Social Behavior and Communication:
Social Groups:

Marmosets form cooperative breeding groups, with shared responsibilities for offspring care and feeding.
Communication:

Vocalizations, body language, and scent marking are key components of marmoset communication.

Reproductive Behavior:

Marmosets exhibit twinning and cooperative breeding, with the entire group participating in caring for infants.

Diet and Feeding Patterns:
Natural Diet:

Marmosets have a varied diet, consuming fruits, insects, tree sap, and small vertebrates.
Captive Diet:

In captivity, marmosets are often fed a balanced diet that mimics their natural nutritional requirements.
Feeding Patterns:

Marmosets have a high metabolic rate, requiring frequent feeding throughout the day.
Conservation and Threats:
Threats to Populations:

Habitat loss, fragmentation, illegal pet trade, and diseases pose significant threats to marmoset populations.

Conservation Efforts:

Conservation initiatives focus on habitat protection, research, education, and community engagement.

Importance of Conservation:

Marmosets play a vital role in maintaining ecosystem health, and conservation efforts contribute to biodiversity preservation.

Human-Marmoset Interaction:

Marmosets as Pets:

Keeping marmosets as pets raises ethical concerns and may contribute to illegal wildlife trade.

Ethical Considerations:

Ethical considerations involve responsible pet ownership, adherence to laws, and considerations for the welfare of marmosets.

Human-Marmoset Conflict:

Habitat encroachment, urbanization, and the pet trade contribute to conflicts, requiring mitigation strategies for coexistence.

Contributions to Science:

Biomedical and Medical Research:

Marmosets serve as valuable models in biomedical research, contributing to studies in neuroscience, genetics, reproductive biology, and pharmacology.

Genetic and Genomic Research:

Their genetic diversity and suitability for genetic engineering contribute to advancements in understanding human health.
Social Behavior and Communication Studies:

Marmosets provide insights into social behavior, communication, and cognition, advancing our understanding of primate intelligence.

Challenges in Research:
Ethical Considerations:

Researchers face ethical challenges related to animal welfare, long-term care, and responsible use of marmosets in research.
Genetic and Biological Variability:

Variability across marmoset populations poses challenges in experimental design and drawing generalizable conclusions.
Technical and Experimental Constraints:

The small size of marmosets presents challenges in certain experimental techniques, surgeries, and behavioral studies.

Reproductive Characteristics:

Twinning and cooperative breeding introduce complexities in reproductive studies and experimental control.

High Costs and Resource Intensity:

The resource-intensive nature of marmoset care and the need for specialized facilities contribute to research challenges.

Marmosets in Art, Literature, Media, and Entertainment:

Art and Literature:

Marmosets have been featured in art and literature, serving as subjects, symbols, and metaphors.

Media and Entertainment:

Marmosets appear in films, television, advertisements, social media, and video games, contributing to popular culture and brand recognition.

Educational Content:

Marmosets are incorporated into educational programs, children's literature, and wildlife workaries to raise awareness about biodiversity and conservation.

Marmosets, with their intriguing biological features, social behaviors, and contributions to science, have left a mark in various aspects of human interaction. From conservation challenges to ethical considerations in research and their portrayal in creative expressions, understanding and appreciating marmosets requires a multidisciplinary perspective that balances scientific inquiry, ethical responsibility, and cultural appreciation.